AI AWARENESS SERIES

AI in Cyber Defense

and Security

Darian Batra

Contents

Introduction

Cybersecurity is no longer just about firewalls and passwords—it's about intelligent systems that can detect, respond to, and even anticipate threats in real time. As attackers become more sophisticated, and as organizations manage growing volumes of data and increasingly complex digital infrastructures, artificial intelligence has become an essential ally in modern cyber defense.

AI in Cyber Defense and Security is part of the AI Awareness Series, designed to bridge the gap between emerging technologies and real-world application. This volume focuses on how AI and machine learning are transforming cybersecurity—from threat detection and incident response to identity management, offensive security, and ethical governance.

Whether you're a cybersecurity professional, IT leader, policymaker, or someone entering the field, this book offers practical insights into how AI is being used to strengthen cyber resilience. It is not a deep dive into algorithms or coding, but a strategic and operational guide for understanding what AI can (and cannot) do in the context of digital defense.

Throughout this book, we'll explore how AI:

- Detects and classifies evolving threats across networks, endpoints, and cloud platforms
- Supports incident response, red teaming, and threat intelligence automation

- Enhances identity and access management, especially in distributed environments
- Handles adversarial attacks and the risks of AI being used offensively
- Powers cybersecurity use cases through generative AI and large language models
- Raises critical questions about ethics, transparency, and regulatory compliance
- Enables organizations to build integrated, AI-enabled security programs

This book combines technical clarity with practical relevance. Each chapter is designed to give you a foundational understanding of the topic, supported by real-world applications and forward-looking insight.

AI isn't a silver bullet—but when applied thoughtfully, it can become a powerful force multiplier in the fight to secure our digital world.

Let's begin.

Chapter 1: Overview of AI and Machine Learning in Security

Let's start by clarifying some important definitions.

Artificial Intelligence refers to technologies that enable machines to perform tasks which usually require human intelligence—like reasoning, problem-solving, and decision-making.

Machine Learning is a subset of AI that focuses on algorithms capable of learning and improving from experience, without being explicitly programmed.

Deep Learning is a more advanced branch of machine learning that uses neural networks with multiple layers, allowing machines to analyze complex data patterns and handle sophisticated tasks.

Finally, Natural Language Processing, or NLP, gives computers the ability to understand, interpret, and even generate human language, which is increasingly important in cybersecurity for tasks like analyzing threat reports or monitoring communications.

Now let's look at how different types of AI are applied specifically in cybersecurity.

Machine Learning is often used for anomaly detection. By identifying patterns and flagging unusual activities, it plays a key role in strengthening cybersecurity defenses.

Deep Learning goes a step further, analyzing complex layers of data to predict and understand sophisticated cyber threats that simpler algorithms might miss.

Natural Language Processing helps security teams make sense of textual information—like threat intelligence reports or intercepted communications—so they can quickly spot potential risks.

To understand where we are today, it's helpful to look at how AI has evolved in the security space.

Early systems relied on rule-based approaches. These systems used predefined rules and signatures to detect known threats—but they struggled with new or evolving attacks.

Modern AI-driven solutions, especially those powered by machine learning, have changed the game. They can identify more complex threats and adapt in real time to new types of attacks.

Chapter 1: Overview of AI and Machine Learning in Security

This evolution has had a major impact on defense strategies, enabling organizations to move from reactive to proactive security, using AI for early threat detection and automated response.

Adopting AI in cybersecurity comes with both significant benefits and serious risks.

On the positive side, AI enables faster and more accurate detection of threats and vulnerabilities, improving overall security posture.

However, AI systems are not foolproof—they can be targeted by adversarial attacks that try to deceive or disable them.

There are also broader concerns about privacy and the potential dangers of relying too heavily on automated decision-making without human oversight.

It's important to balance the advantages of AI with awareness of its limitations and risks.

To wrap up, AI and machine learning are transforming cybersecurity, offering powerful tools for threat detection, analysis, and response.

But with these advances come new challenges—especially around privacy, security, and reliance on automation.

Understanding both the potential and the pitfalls of AI is critical for anyone working in or studying cybersecurity today.

Chapter 2: Threat Landscape and Attack Vectors

Cyber threats today are becoming increasingly complex. Attackers are no longer relying on simple methods; instead, they use advanced techniques specifically designed to bypass traditional security defenses. Automation and AI allow them to scale attacks rapidly and exploit vulnerabilities more efficiently than ever before. This constant evolution of tactics challenges even the most robust cybersecurity defenses, making it harder for organizations to keep up.

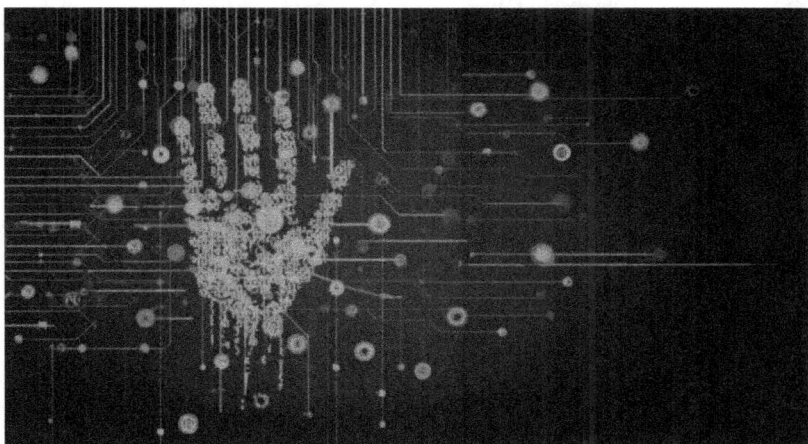

AI has expanded the attack surface by enabling automated reconnaissance—this means attackers can gather vast amounts of data at incredible speeds, identifying potential targets with minimal effort. AI also allows for adaptive exploits, where attacks evolve in real-time to bypass defenses. On top of that, social engineering has become more convincing with AI's ability to craft sophisticated phishing and manipulation techniques.

Chapter 2: Threat Landscape and Attack Vectors

Traditional attack vectors like malware and phishing have also evolved due to AI. Malware can now adapt and change its behavior to avoid detection. Phishing attacks, once generic, are now highly targeted—AI can mimic legitimate communications so well that even trained users might be fooled. This evolution makes defending against these attacks significantly more challenging.

Let's look at how AI has enhanced malware. Modern AI-driven malware learns from its environment, adapting to avoid detection and remain hidden longer. Its payload can change dynamically, making traditional defenses less effective. Moreover, AI optimizes the attack strategy itself, ensuring maximum impact by selecting the most efficient pathways for exploitation.

There have been several real-world incidents involving AI-driven ransomware. Attackers use AI to automate phishing attacks, exploit vulnerabilities, and evade detection throughout their campaigns. These tactics make ransomware harder to detect and stop, and they force defenders to rethink their strategies. Combating AI-powered ransomware requires advanced, adaptable defenses that can keep up with the rapidly changing tactics of attackers.

To counter AI-powered threats, organizations are adopting AI-based detection systems that offer real-time threat identification. Behavioral analytics is also crucial—it allows security teams to spot unusual patterns that might indicate a threat. Finally, a multi-layered security approach ensures that even if one defense fails, others can help protect the organization from diverse AI-driven attacks.

Chapter 2: Threat Landscape and Attack Vectors

Adversarial AI is a concept where attackers craft deceptive inputs designed to fool machine learning models. These attacks target vulnerabilities in AI systems, manipulating inputs so that the AI makes incorrect decisions—like misclassifying malicious activity as benign. This introduces a whole new category of risks for cybersecurity teams to address.

Attackers use several techniques to evade AI-based detection systems. One method is poisoning training data, which degrades the AI's performance. They also create adversarial examples—inputs that look normal but are specifically designed to mislead the AI. Additionally, they mimic legitimate behavior patterns, making it even harder for anomaly detection systems to flag their activities.

There are real-world cases where adversarial AI has been used successfully against cybersecurity defenses. These incidents highlight how attackers can manipulate AI systems to bypass protections. Such examples underline the urgent need to reinforce AI models against adversarial threats and ensure they're robust enough to handle these kinds of attacks.

AI is changing phishing attacks by making them more believable and harder to detect. Attackers use AI-generated content to create phishing messages that closely mimic legitimate communications. They also automate the distribution of these phishing campaigns, targeting a vast number of victims quickly. This realistic approach increases the chance that victims will engage with the malicious content.

Machine learning allows attackers to analyze massive datasets to find patterns and vulnerabilities in individuals. They use this information to craft highly personalized social engineering attacks. By tailoring their approach to each target's specific weaknesses, attackers significantly increase their chances of success.

Defending against AI-driven social attacks requires a combination of strategies. First, user education is vital—helping people recognize phishing and social engineering tactics. AI-based detection tools can also block many of these threats before they reach the user. Behavioral analytics further enhance security by spotting unusual activity. Finally, strong authentication measures provide an essential layer of protection against account compromise.

Deepfakes are synthetic media—audio or video—that convincingly imitate real people using AI. They present serious challenges for verifying the authenticity of content, making it harder to trust what we see and hear. This erosion of trust is a significant concern in both cybersecurity and public information.

Deepfakes are increasingly used for fraud and impersonation. Attackers can convincingly impersonate executives, tricking employees and stakeholders. They also use deepfakes to manipulate victims into sharing sensitive information or transferring money. On a broader scale, deepfakes contribute to the spread of misinformation, damaging trust in media and information sources.

To counter deepfake threats, organizations are developing detection technologies and countermeasures. These tools analyze media for signs of manipulation, helping identify and stop deepfake-based attacks. As the technology behind deepfakes evolves, so must the detection methods we use to safeguard authenticity and trust.

To wrap up, AI is fundamentally reshaping the cybersecurity landscape—both in terms of threats and defenses. Attackers leverage AI for more sophisticated, scalable, and adaptive attacks, making traditional defenses less effective. Understanding these evolving tactics is crucial for building effective defense strategies. As AI continues to advance, staying informed and adaptive will be key to maintaining cybersecurity resilience.

Chapter 3: Core AI Techniques for Cyber Defense

Artificial Intelligence is revolutionizing cybersecurity. First, AI enhances threat detection by analyzing data for patterns and anomalies, helping to identify complex attacks faster. Second, AI enables automated responses to incidents, reducing reaction time and minimizing damage. Finally, AI analyzes massive volumes of security data, identifying new threats and vulnerabilities before they can be exploited.

There are several key benefits to integrating AI into cyber defense strategies. AI increases detection speed, allowing faster responses. It improves accuracy, reducing false positives and better identifying real threats. AI also scales efficiently, monitoring large networks with ease. It supports proactive defense by predicting and preventing attacks. And importantly, it reduces manual effort by automating monitoring and response tasks.

However, AI-based solutions also come with challenges. The quality of data is critical—poor data can cause inaccurate results. AI models are susceptible to adversarial attacks that trick the system. There are also concerns about interpretability—understanding AI decisions can be difficult. Lastly, AI systems need continuous learning to adapt to new threats and remain effective.

Anomaly detection is crucial in cybersecurity. It works by identifying patterns that deviate from normal behavior, helping to uncover security threats that might otherwise go unnoticed. By spotting these anomalies, organizations can detect breaches or insider threats early.

Behavioral analytics supports anomaly detection by profiling user and system behavior. It learns typical activity patterns and then monitors for suspicious or unusual actions that deviate from these norms. This profiling helps catch unauthorized activities quickly.

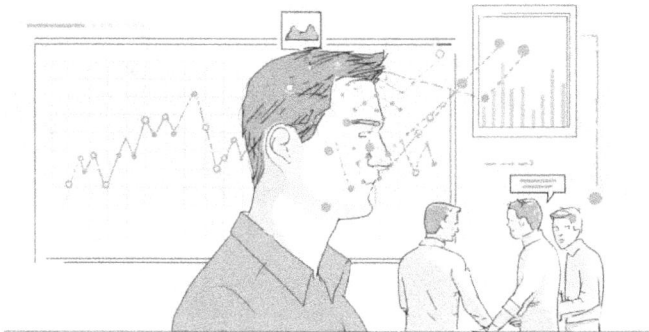

Real-world case studies highlight how effective anomaly detection and behavioral analytics can be. They've been successful in uncovering advanced threats and malicious behavior patterns. When used together, these techniques help neutralize threats before they cause significant harm.

Pattern recognition plays a key role in cyber defense by identifying known threat signatures. It helps detect recurring behaviors or attack patterns, enabling faster response times and improved security posture.

Clustering algorithms are used to group similar threats together. This categorization helps prioritize risks, making it easier for security teams to focus on the most critical threats first and develop targeted defense strategies.

In practice, pattern recognition enhances intrusion detection by spotting suspicious behaviors in network traffic. Clustering is particularly useful in malware analysis—it groups similar malware samples, speeding up threat analysis and improving automated threat classification.

Chapter 3: Core AI Techniques for Cyber Defense

Natural Language Processing, or NLP, plays a key role in extracting threat intelligence. It processes unstructured data from security logs and reports, turning it into structured information that can be analyzed effectively. NLP also helps identify new threats by detecting patterns in large datasets.

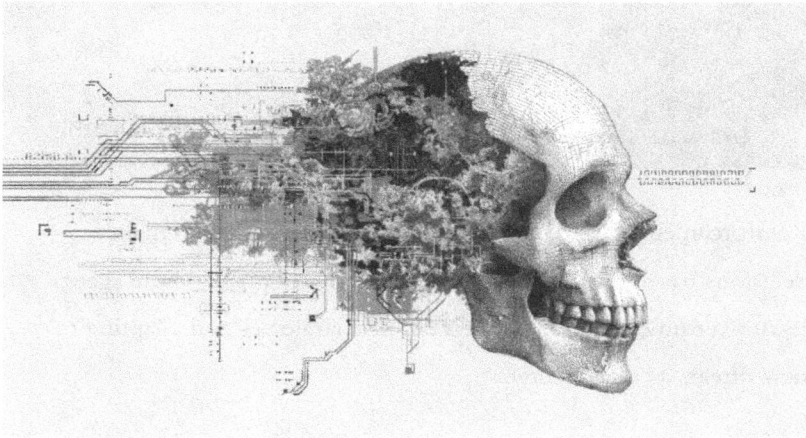

Several NLP techniques are used in cybersecurity. Entity recognition identifies important terms and entities in text. Sentiment analysis examines tone and intent, which can help detect threats. Topic modeling uncovers hidden themes in security data, aiding in strategic threat assessments.

Real-world applications of NLP include automating threat report classification and summarizing complex threat intelligence. These techniques speed up understanding and enable quicker responses to new threats, enhancing overall security operations.

Reinforcement learning, or RL, allows security agents to make decisions based on feedback from their environment. These agents learn over time, optimizing their defense strategies and adapting to new threats as they emerge.

RL powers adaptive defenses by enabling systems to update their security policies dynamically. This allows for real-time responses to new attack techniques and improves system resilience by learning from past incidents.

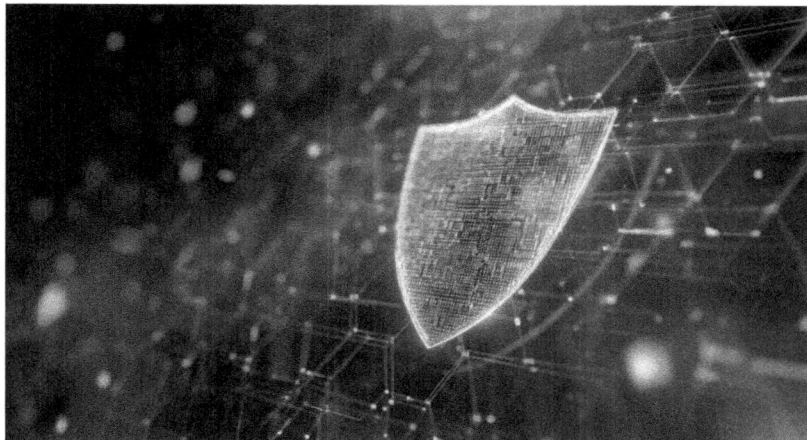

Looking ahead, reinforcement learning will drive the development of autonomous, self-improving cyber defense systems. These systems will reduce the need for human intervention, learn from past attacks, and proactively anticipate and prevent future threats.

To conclude, AI techniques like anomaly detection, pattern recognition, NLP, and reinforcement learning are transforming cyber defense. While there are challenges, the benefits—faster detection, automated responses, and adaptive defenses—make AI an essential tool for protecting against evolving cyber threats.

Chapter 4: Data and Infrastructure Requirements

AI systems depend on various types of data. They need both structured data, like databases, and unstructured data, such as images or text. For supervised learning, labeled datasets are critical because they help models learn correct patterns. Beyond that, continuous data streams allow AI systems to stay updated and responsive, especially in applications that need real-time feedback.

Infrastructure plays a huge role in AI. High-performance computing resources, like GPUs and specialized chips, are needed for heavy training tasks. Storage systems must be scalable to handle the massive amounts of data AI consumes. Finally, robust networking is key — it connects computing resources and storage, allowing smooth and efficient data flow throughout AI workflows.

Data quality and infrastructure are tightly linked. Without reliable infrastructure, data collection, storage, and processing become unreliable too. Poor infrastructure can lead to data loss or corruption, harming data integrity. In fact, good data and strong infrastructure go hand in hand — you need both for successful AI outcomes.

When collecting data, there are several common challenges. Privacy concerns can limit access to sensitive data, making it hard to get comprehensive datasets. Data scarcity is another issue — sometimes, the data you need simply isn't available. And sampling bias means that if you're not careful, your datasets might not represent the real world, which affects how well your AI models perform.

Data annotation is another critical area with its own complexities. Skilled annotators are essential for high-quality labeling, and standardized protocols make sure the work is consistent. Quality control measures are important to catch mistakes early. And while automation tools can help speed things up, they should work alongside human efforts to ensure accuracy.

Bias in datasets can lead to unfair or harmful AI decisions. That's why it's vital to source data from diverse and representative places. Bias detection techniques help spot and address issues before they affect your models. And continuous monitoring ensures that your data — and your AI systems — stay fair, accurate, and ethical over time.

Chapter 4: Data and Infrastructure Requirements

Protecting privacy is a big concern in AI. Techniques like federated learning allow models to train on decentralized data, so sensitive information doesn't leave local devices. Differential privacy works by adding random noise to data or outputs, making it hard to trace results back to individuals. These techniques help keep personal data safe while still enabling AI development.

Balancing privacy with model accuracy can be tricky. More privacy controls often mean less data to learn from, which can lower accuracy. It's important to find the right balance between protecting user data and maintaining effective AI performance.

AI systems also need to comply with data privacy laws like GDPR and CCPA. That means designing systems with these laws in mind from the very beginning. Compliance isn't just about avoiding penalties — it's about respecting user rights and building trust with the people your AI serves.

Chapter 4: Data and Infrastructure Requirements

Choosing between cloud and on-premises deployment comes down to trade-offs. Cloud platforms offer great scalability and flexibility, letting you adjust resources as needed and pay only for what you use. On the other hand, on-premises solutions might save money in the long run but come with high upfront costs and require regular maintenance.

Security and compliance also differ between cloud and on-premises deployments. With on-premises, you have direct control over data security and compliance. Cloud providers offer strong security tools, but you're trusting a third party to manage your data, which comes with its own risks.

Hybrid approaches can offer the best of both worlds. By combining cloud scalability with on-premises control, you can tailor your setup to your needs. For example, you might keep sensitive data on-premises for security while using the cloud for general computing tasks. The decision really depends on your workload, security requirements, and budget.

Training AI models starts with choosing the right algorithms and preparing your datasets carefully. Clean, high-quality data leads to better model performance. You'll also need to plan your resources — making sure you have enough computing power is key to efficient training.

Continuous learning allows models to improve over time by learning from new data. Automated retraining pipelines help update models regularly without needing manual input, keeping performance high. This approach also helps prevent model degradation — when models become less accurate as the world around them changes.

Monitoring AI models after deployment is crucial. By watching their performance, you can catch issues early and retrain models before problems grow. Managing the model lifecycle means keeping them accurate, compliant, and effective at every stage — from initial deployment through updates and beyond.

To wrap up, developing AI solutions requires a balanced focus on both data and infrastructure. You need diverse, high-quality data and the right infrastructure to support collection, storage, and processing. Addressing privacy, deployment, and retraining strategies ensures your AI remains effective, fair, and trustworthy throughout its lifecycle.

Chapter 5: Network and Endpoint Security

Cyber threats are constantly evolving. Attackers develop new techniques to bypass security measures and exploit vulnerabilities we may not have considered. These threats don't just target networks — they also go after endpoints like laptops and mobile devices. That's why it's critical to understand how threats evolve, so we can design defenses that adapt and protect effectively.

Let's look at how network and endpoint security differ and where they overlap.

Network security focuses on guarding the perimeter and monitoring traffic to prevent unauthorized access.

Endpoint security, on the other hand, protects individual devices from malware and direct attacks.

But when you integrate both approaches, you get a much stronger defense — they work together to cover gaps the other might miss.

To build a truly secure environment, you need a layered defense strategy.

That means implementing multiple security layers, maintaining consistent policies across all systems, and ensuring you're applying regular updates and patches.

It's also essential to have real-time threat monitoring in place, so you can detect and respond to incidents as they happen.

Intrusion Detection and Prevention Systems — or IDPS — rely on a few core technologies.

Signature-based detection compares network activity against known threat signatures.

Chapter 5: Network and Endpoint Security

Anomaly-based detection looks for behavior that deviates from the norm, flagging potential intrusions.

And hybrid systems combine both methods to improve accuracy and reduce false alarms.

Where and how you deploy IDPS matters.

Your network's topology will influence deployment choices to ensure effective monitoring.

High traffic environments need scalable architectures to handle the load without sacrificing performance.

And you can choose between inline or passive monitoring setups, depending on your security goals.

Despite their strengths, IDPS solutions face challenges in real-world use.

Chapter 5: Network and Endpoint Security

False positives can overwhelm security teams with unnecessary alerts.

Encrypted traffic often hides potential threats from detection.

And resource limitations can hamper an IDPS's ability to process and analyze data in real time.

Artificial Intelligence plays a growing role in threat detection.

AI can analyze behavior to spot suspicious activity as it happens.

It uses pattern recognition to detect known threats within massive datasets.

And with predictive analytics, AI can even forecast likely threats before they emerge, giving security teams a crucial head start.

Modern antivirus and EDR — or Endpoint Detection and Response — tools bring several powerful capabilities.

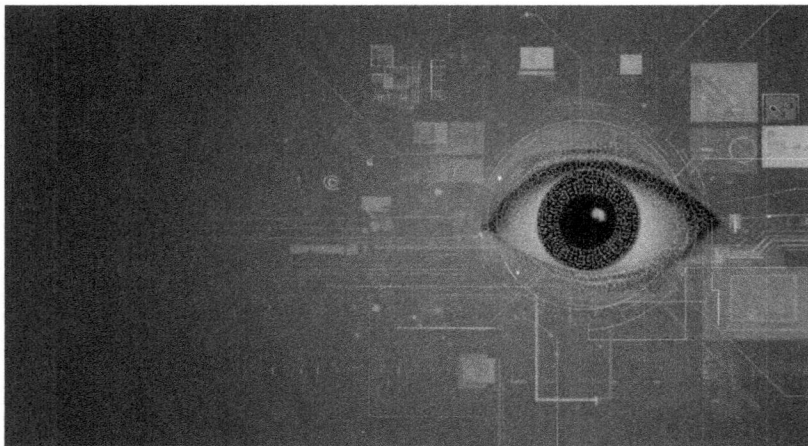

AI-enhanced detection makes them faster and more accurate.

Chapter 5: Network and Endpoint Security

They offer real-time monitoring of endpoints, catching threats as they occur.

With automated response, these tools can act immediately to neutralize threats.

And their comprehensive visibility across all devices simplifies security management.

Real-world case studies show the impact of AI-driven security tools.

Organizations using these tools report significant reductions in security breaches.

They also experience faster incident response times, meaning threats are identified and dealt with much quicker — reducing the risk of serious damage.

Now, let's look at Zero Trust — a model built on the idea that nothing is trusted by default.

This means every access request must be verified, even from inside the network.

Access is granted based on the least privilege principle, reducing exposure to risks.

And micro-segmentation helps isolate critical assets, preventing attackers from moving freely within the network.

Chapter 5: Network and Endpoint Security

Continuous verification is a key part of Zero Trust in enterprise environments.

It means constantly checking the identity and security posture of users and devices, not just at login but throughout their session.

Zero Trust offers strong security benefits but also comes with challenges.

It improves overall security by enforcing strict access controls and continuous verification.

However, it often requires a cultural shift — organizations need to foster security awareness at all levels.

Integration can be complex, especially when dealing with legacy systems.

And adopting Zero Trust usually means investing in advanced technologies and continuous monitoring capabilities.

Chapter 5: Network and Endpoint Security

To wrap up, we've explored how modern security strategies — from integrated network and endpoint protection to AI-driven tools and Zero Trust models — help safeguard enterprise environments.

The key takeaway is that no single solution is enough on its own. A combination of smart technologies, consistent policies, and vigilant monitoring is essential for staying ahead of evolving threats.

Chapter 6: Threat Intelligence and Incident Response

Threat intelligence is all about gathering and analyzing information on cyber threats to strengthen our defenses. Meanwhile, incident response is the structured approach we use to handle security breaches—keeping damage to a minimum and restoring operations as quickly as possible.

Today's cybersecurity threats are becoming more advanced, with persistent threats and zero-day exploits making headlines. Because of this, organizations must constantly adapt both their threat intelligence efforts and their response strategies to stay ahead of evolving cyber risks.

These processes—threat intelligence and incident response—are critical for any organization's security. They help detect threats proactively, enable quick action when breaches occur, and ultimately help maintain business continuity by minimizing disruption and protecting key assets.

Automated threat hunting focuses on proactively detecting threats before they cause harm. By analyzing patterns in data automatically, organizations can spot suspicious activity early. This allows security teams to devote more time to investigating complex threats, where human expertise really counts.

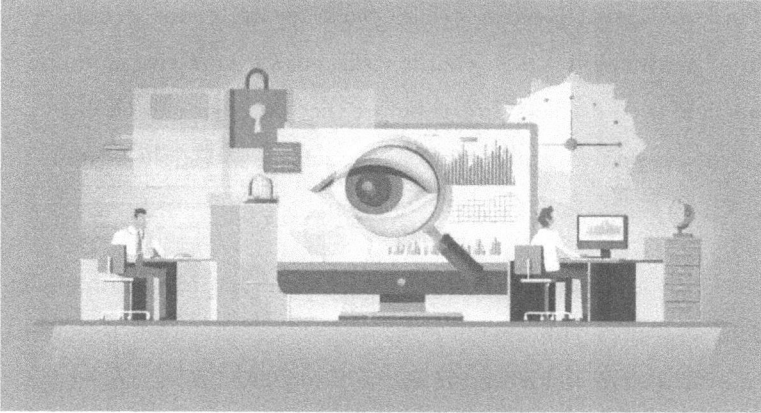

Key tools in automated threat hunting include SIEM systems, which collect and analyze security data, EDR tools that monitor endpoints in real time, and machine learning algorithms that help detect anomalies and predict potential threats before they escalate.

Automation brings big advantages to threat hunting, like faster detection and fewer human errors. But it's not without challenges—false positives can still happen, and automated systems need constant tuning by experts to stay effective and accurate.

Chapter 6: Threat Intelligence and Incident Response

Artificial intelligence is changing how Security Operations Centers work. AI automates routine security tasks, analyzes massive amounts of data, and even predicts threats before they happen. This allows security teams to focus on investigating complex and sophisticated attacks.

With AI, SOCs can detect anomalies in real time and trigger automated incident responses. However, human analysts still play a key role, using AI-generated alerts to investigate and respond with greater speed and accuracy.

In the real world, AI-driven incident response has helped organizations identify and respond to threats faster, reducing downtime and damage. AI also improves early detection of advanced threats, strengthening security posture and boosting operational efficiency.

SOAR platforms help centralize and unify incident management. By automating workflows, SOAR reduces the manual workload and speeds up the entire threat response process. It also improves coordination across different security tools and teams, making operations more effective.

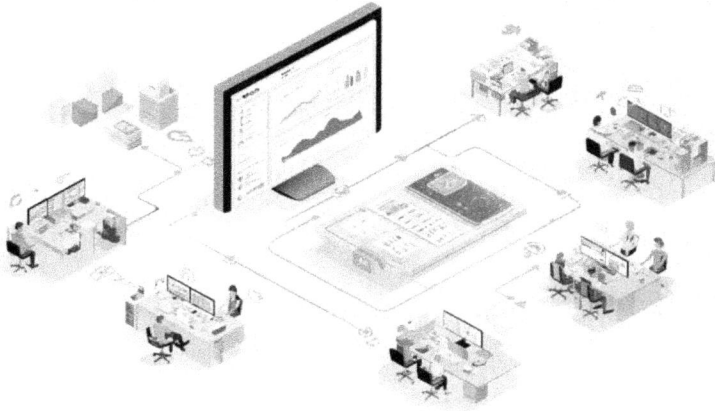

SOAR platforms offer several powerful features—automated playbooks for response, case management tools for tracking incidents, seamless integration with threat intelligence sources, and dashboards that provide clear insights into security operations.

SOAR integrates with key security tools like SIEM and EDR systems. It also works alongside firewalls and external threat feeds, helping automate security workflows and enhance overall protection through better coordination and data sharing.

Predictive threat intelligence uses data analysis to forecast future attacks. By examining both historical and real-time data, organizations can detect threats earlier and even prevent incidents from occurring, making their cybersecurity defenses far more proactive.

The data for predictive intelligence comes from a variety of sources—network logs, dark web monitoring, malware databases, and global threat feeds. Advanced analytics, including machine learning, are applied to this data to generate actionable predictions.

By anticipating threats, organizations can identify vulnerabilities before attackers exploit them. This means they can prioritize resources effectively, strengthen defenses in key areas, and reduce their overall exposure to cyber risks—helping to ensure business continuity.

To wrap up, we've explored how modern approaches in threat intelligence and incident response—especially with the help of automation, AI, SOAR platforms, and predictive analytics—are transforming cybersecurity. By applying these advanced methods, organizations can stay ahead of threats, minimize risk, and protect their operations in an increasingly complex digital world.

Chapter 7: Identity and Access Management

Identity and access management, or IAM, are built on three core principles. First is identity verification — confirming users are who they say they are before granting access. Second is access rights management — controlling who can see or modify sensitive resources. And finally, authorization enforcement — making sure only approved users access protected systems, which helps preserve overall security.

Traditional IAM systems face several challenges. Scalability issues mean they struggle to handle growing numbers of users and systems. Complex user management adds to the risk of errors, especially in large organizations with many roles. And lastly, security vulnerabilities — outdated authentication methods and insider threats make these systems prone to breaches.

IAM plays a vital role in modern cybersecurity. It ensures secure user access by allowing only verified users into systems. Through policy enforcement, it applies consistent security measures across the organization. And with AI integration, IAM systems can detect and respond to threats in real time, adapting dynamically to emerging risks.

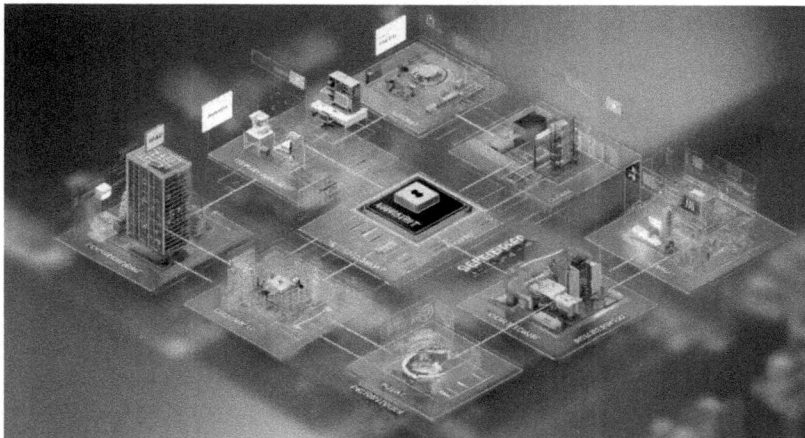

Biometric authentication uses unique biological characteristics for security. Fingerprint scanning is reliable and widely adopted. Facial recognition analyzes facial features and is common in smartphones and surveillance. And voice recognition authenticates users based on distinct vocal patterns, adding another layer of security in many applications.

Artificial intelligence significantly boosts biometric systems. It allows for advanced pattern recognition, leading to more accurate identifications. AI also helps in reducing errors like false positives and negatives, making systems more reliable. And thanks to adaptive learning, AI systems stay effective by adjusting to changes in biometric data over time.

Biometrics offer strong security benefits due to their unique identifiers, but they come with concerns. Privacy issues arise around how biometric data is stored and whether users give informed consent.

Chapter 7: Identity and Access Management

From an ethical perspective, organizations must guard against misuse of biometric data and commit to clear, transparent policies.

Multi-factor authentication, or MFA, has evolved over time. It started with static factor combinations like passwords paired with tokens. Today, we see context-sensitive methods that adapt based on user behavior and context. AI and behavioral analytics now enhance security by assessing risk during authentication, making access decisions smarter.

Adaptive authentication uses context-aware approaches, changing authentication requirements based on factors like location, device, or behavior. This flexibility helps improve both security and user experience by adjusting the security process dynamically according to the situation.

Risk-based access control involves a continuous evaluation of threats. By analyzing user behavior over time, systems can identify suspicious activity early. And with an immediate response mechanism, these systems can take action instantly when a potential threat is detected, minimizing risks to the organization.

Insider threats come from individuals within the organization — trusted employees or users — who may either deliberately or accidentally compromise security. These threats are particularly dangerous because insiders often have privileged access to sensitive information and critical systems.

IAM can help detect suspicious behavior by monitoring user actions and looking for anomalies. Techniques like behavior tracking and access pattern analysis enable organizations to spot potential threats before they escalate into actual breaches.

AI-driven analytics enhance insider threat detection by continuously learning from patterns in user data. This allows for proactive detection of unusual or risky behavior. By identifying potential threats early, organizations can respond quickly and prevent security incidents before they happen.

Chapter 7: Identity and Access Management

To wrap up, identity and access management is a critical part of cybersecurity, especially when enhanced with AI-driven technologies. From verifying identities and managing access to detecting insider threats, IAM solutions help protect organizations in an increasingly digital and complex world.

Chapter 8: Cloud and Application Security

Cloud and application security has become a top priority as organizations increasingly rely on cloud services. As cloud adoption rises, so does the risk of cyber threats. Sophisticated attacks require continuous updates to security strategies. Effective cloud security is essential for keeping businesses running smoothly and maintaining customer trust.

Let's look at some common threats in cloud environments. Data breaches can cause huge financial and reputational losses. Insecure APIs are a major risk, often exploited by attackers. Misconfigurations are another critical issue—simple mistakes can open the door to attacks. And we can't forget insider threats, whether intentional or accidental, they pose a serious risk to security.

AI plays a crucial role in boosting security. It can detect threats in real time, allowing for immediate action. AI also automates responses, which reduces the time it takes to react and limits damage. Using predictive analytics, AI spots patterns and anomalies that traditional tools might miss, making it a powerful asset in defending against attacks.

Cloud misconfigurations happen when cloud settings don't follow security best practices, which leaves systems exposed. These errors can let unauthorized users access sensitive data, leading to breaches. The impact can be severecausing data leaks, service outages, and even regulatory fines for non-compliance.

AI can help identify and fix misconfigurations. Using anomaly detection, machine learning models scan configuration data for unusual patterns that might signal a security risk. Automated remediation tools can fix these problems quickly—much faster and more efficiently than doing them manually.

There are real-world examples of how AI-driven tools have successfully detected misconfigurations before they led to breaches.

These case studies show how organizations can use AI solutions to strengthen their cloud security posture and prevent costly incidents.

In DevSecOps, integrating AI into the DevOps pipeline is key for automating security. AI can perform automated security checks, helping detect vulnerabilities early in the development cycle. It also speeds up vulnerability detection, ensuring secure software delivery without slowing down the process.

AI-driven tools analyze codebases for security flaws, offering suggestions on how to fix them. They help prioritize risks based on severity so teams can tackle the most critical issues first. This improves workflows in DevSecOps, making code more secure and reducing the time needed for remediation.

AI brings great benefits to DevSecOps by improving efficiency and accuracy. However, there are challenges too. Integrating AI tools into existing pipelines can be complex. False positives are another risk, as they can create noise and disrupt workflows. Plus, organizations need skilled personnel who understand both AI and security to manage these tools effectively.

APIs and microservices present unique security challenges. Weak authentication can expose systems to unauthorized access. If APIs aren't secured properly, sensitive data might be leaked. And with microservices, the increased number of distributed endpoints creates a larger attack surface, making security even more critical.

AI helps monitor and protect APIs by detecting unusual behavior and spotting potential threats early. It also prevents abuse by analyzing API usage patterns and blocking suspicious activities in real time. Additionally, AI can adapt its threat responses dynamically, staying ahead of evolving threats.

To protect microservice environments, continuous monitoring is a must. Using behavioral analytics allows AI to detect anomalies and breaches more effectively. It's also important to make sure AI tools work seamlessly with existing security frameworks so they enhance, rather than complicate, the overall security architecture.

To wrap up, AI is a game changer in cloud and application security. It enhances detection, automates responses, and supports secure development practices. However, organizations need to be mindful of integration challenges, potential false positives, and the need for skilled professionals. By balancing these factors, AI can be a powerful ally in building stronger, more resilient security systems.

Chapter 9: AI in Offensive Security (Red Teaming)

Red teaming plays a crucial role in cybersecurity by simulating real-world attackers in a controlled environment. These ethical hacking exercises help organizations uncover hidden vulnerabilities before malicious actors can exploit them. The findings from red team operations allow defenders to bolster their security strategies and prepare for actual threats more effectively.

Artificial Intelligence enhances offensive security by automating attack methods, making them faster and more scalable. AI can analyze large volumes of data to improve decision-making during attacks, quickly identifying vulnerabilities and suitable attack paths. It also allows attackers to adapt their methods on the fly, evading defenses in real-time. With AI, red teams can conduct operations on a much larger scale than with traditional, manual approaches.

Despite the advantages AI brings to offensive security, there are significant ethical concerns. The same tools that help identify vulnerabilities could be misused, raising questions about responsible usage. Privacy is a major issue, as AI-driven attacks may access sensitive data. Therefore, it's essential that AI use in security is guided by strict oversight and clear ethical guidelines to prevent harm.

Automated penetration testing leverages AI to perform security testing without the need for constant human input. This automation significantly speeds up testing cycles, enabling quicker discovery of vulnerabilities. Automated tools also allow for continuous monitoring, keeping defenses active around the clock. And by reducing the reliance on human testers, organizations can lower their labor costs associated with manual testing efforts.

Traditional penetration testing is heavily dependent on human expertise and manual effort, which limits its scalability and speed. In contrast, AI-driven testing offers quicker, more adaptable methods that can cover a wider attack surface. However, it's important to recognize that AI tools may still lack the deep contextual understanding that skilled human testers bring, especially in complex environments.

Generative AI models are designed to analyze massive datasets of known exploits and vulnerabilities, learning how they function. These models can then create entirely new code variants that are capable of bypassing existing security defenses. This capability introduces both exciting possibilities and serious risks for cybersecurity.

Chapter 9: AI in Offensive Security (Red Teaming)

Real-world examples have shown that generative AI can be used to craft working exploits that challenge traditional security measures. This forces security teams to enhance their analysis and response times to emerging threats. While AI innovation drives the field forward, it also presents new security risks that require continuous vigilance and proactive mitigation strategies.

While AI-generated exploits offer new offensive capabilities, they also come with inherent risks and limitations. These include the potential for unintended misuse, the difficulty in predicting how these exploits might evolve, and the challenge of ensuring that AI-generated tools are used ethically and responsibly within legal boundaries.

AI-powered reconnaissance leverages diverse data sources — from social media to network traffic — to gather information much faster than traditional manual methods. This rapid data processing uncovers insights that might otherwise be overlooked, giving offensive teams a valuable edge during the early stages of an attack campaign.

Chapter 9: AI in Offensive Security (Red Teaming)

Machine learning models can process gathered data to classify and prioritize potential targets based on risk factors and attack surface. This targeted approach allows offensive teams to focus their efforts more efficiently, improving the effectiveness of their campaigns and maximizing their impact on high-risk vulnerabilities.

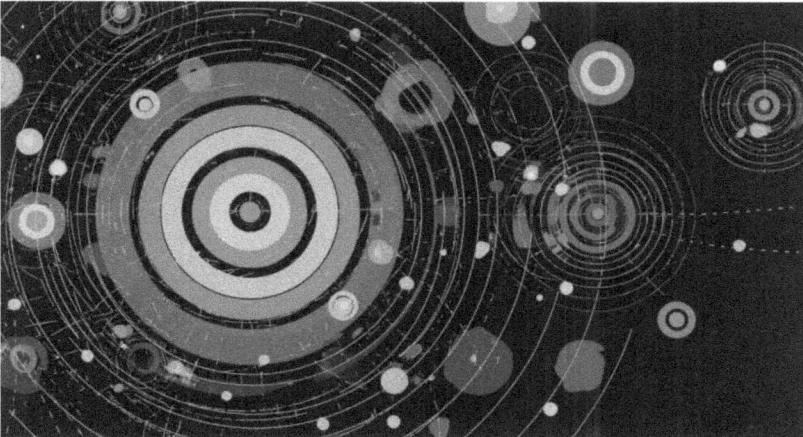

AI also plays a significant role in social engineering attacks like phishing. By personalizing phishing messages based on collected data, AI makes these attacks more convincing and harder to detect. It can also identify individuals more likely to fall for such attacks, increasing the success rates of offensive campaigns that rely on manipulation and deception.

To wrap up, AI has transformed offensive security by enhancing the speed, scale, and effectiveness of red teaming operations. While it brings significant advantages in areas like automation, exploit generation, and reconnaissance, it also introduces serious ethical and security risks. It's critical for organizations and professionals to

Chapter 9: AI in Offensive Security (Red Teaming)

approach AI in offensive security with a balance of innovation, caution, and a strong ethical framework.

Chapter 10: Countering Adversarial AI

Adversarial AI refers to inputs that are deliberately designed to mislead AI models. These inputs exploit weaknesses, causing incorrect decisions or predictions. They can degrade model performance, making AI systems less accurate and less trustworthy. Even more concerning, adversarial attacks can create security risks—potentially leading to breaches that expose vulnerabilities in AI-powered systems.

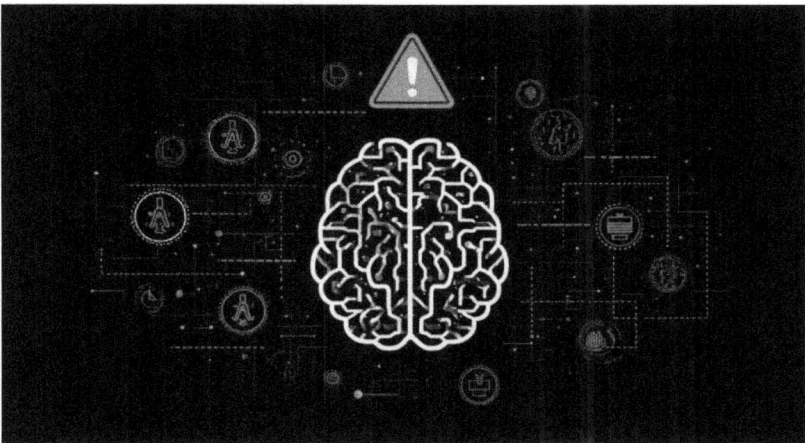

Securing AI systems is critical because adversarial attacks can have serious consequences. When AI models are compromised, it's not just the predictions that suffer—real-world outcomes and security can be at stake. By understanding and countering these attacks, we help ensure AI systems remain reliable, trustworthy, and secure.

There are several approaches to defending against adversarial attacks.

Chapter 10: Countering Adversarial AI

First, adversarial training exposes AI models to malicious examples during training, making them more robust.

Second, input preprocessing techniques help sanitize and filter inputs before they reach the model.

Third, detection mechanisms work in real time to spot and flag adversarial attacks.

Finally, architectural changes involve modifying the structure of models to improve their resilience to attacks.

Adversarial examples can be created using different methods.

Gradient-based methods use information from the model's gradients to make small, targeted changes that trick the model.

Optimization-based approaches fine-tune inputs to maximize the error while keeping changes subtle.

And heuristic algorithms apply rule-based tweaks without relying on model gradients, making them useful against a wide range of systems.

Let's look at some common attack strategies.

The Fast Gradient Sign Method, or FGSM, creates small perturbations designed to deceive models with minimal changes.

Projected Gradient Descent, or PGD, builds on FGSM by applying multiple, iterative perturbations while staying within defined limits, making it a stronger attack.

Chapter 10: Countering Adversarial AI

Adversarial examples play a big role in AI security research.

They help identify vulnerabilities in AI models, revealing weak spots before attackers do.

They also aid in developing more robust defenses, helping researchers improve model security and reliability.

And by simulating real-world attacks, they allow for realistic testing of AI system defenses.

Measuring how well a model withstands adversarial attacks is crucial.

Robustness assessment involves checking if a model can handle maliciously crafted inputs.

By testing with adversarial examples, we can find weaknesses that might otherwise go unnoticed, improving overall security.

Chapter 10: Countering Adversarial AI

To evaluate robustness effectively, we use several benchmarks and metrics.

Accuracy under attack measures how well a model performs when faced with adversarial inputs.

Perturbation norms quantify how much the input has been changed.

Robustness scores give numerical indicators of how resilient a model is.

And standardized benchmarks ensure that tests are consistent and repeatable across different models and scenarios.

Several tools help with robustness and resilience testing.

The CleverHans framework provides libraries for generating adversarial attacks.

Foolbox offers tools to craft adversarial examples and evaluate models.

And the ART framework gives a comprehensive set of tools for both generating attacks and assessing defenses in AI systems.

Defensive distillation is a technique designed to make AI models less sensitive to adversarial inputs.

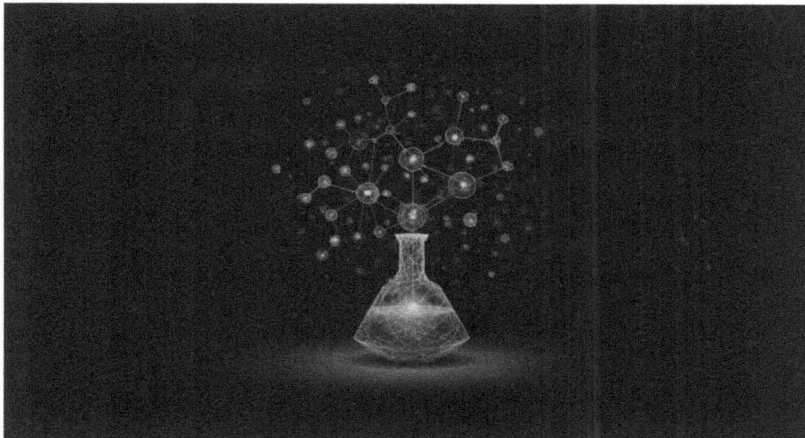

It works by training models in a way that smooths out their decision boundaries, reducing their vulnerability to small input changes crafted by attackers.

There are several techniques for hardening models against attacks.

Adversarial training, which we've mentioned before, involves training models with adversarial examples.

Gradient masking hides gradient information to make it harder for attackers to craft effective inputs.

Feature squeezing simplifies inputs to minimize the impact of perturbations.

And model ensemble approaches combine several models, boosting overall resilience by making it harder for attacks to fool every model at once.

No single defense method is foolproof.

Chapter 10: Countering Adversarial AI

Adversaries are constantly developing new tactics, so relying on one approach leaves gaps.

A layered defense strategy—using multiple defenses together—helps address a broader range of attacks.

And continuous evaluation is key to keeping up with new threats, ensuring that defenses stay effective over time.

To wrap up, countering adversarial AI requires a combination of strategies.

From understanding the nature of attacks to testing model robustness and applying defensive techniques, every layer of defense counts.

By staying proactive and continuously updating our defenses, we can build AI systems that are more secure, reliable, and resilient against adversarial threats.

Chapter 11: Ethics and Responsible AI in Cybersecurity

Let's start by looking at the core ethical principles in AI-driven cybersecurity.

First, Fairness in AI—we want to make sure AI systems don't introduce bias and treat all users equitably.

Next, Transparency Standards—AI decisions shouldn't feel like a "black box." They need to be explainable so users can trust them.

Privacy Protection is critical—AI systems must safeguard personal data and protect user privacy.

And finally, Security Assurance—AI itself must be resilient against cyber threats and designed with security in mind.

Why is responsible AI deployment so important in cybersecurity?

First, it's about Enhancing Cybersecurity Safely—AI should strengthen security without causing unintended harm.

Ethical AI Design ensures that systems meet ethical standards right from development through to deployment.

And Monitoring the AI Lifecycle means we're continuously checking AI performance and addressing issues before they create risks.

Ethical AI in cybersecurity faces both challenges and opportunities.

Chapter 11: Ethics and Responsible AI in Cybersecurity

We have Evolving Cyber Threats—AI must adapt quickly to keep up with new attack methods.

There are Data Privacy Concerns—protecting user data remains a huge challenge.

But there are also Opportunities to Improve Security—ethical AI can enhance detection accuracy and help reduce bias, leading to better outcomes for everyone.

Now let's talk about bias and fairness in security algorithms.

One major source of bias is Unrepresentative Training Data—if the data doesn't reflect the real world, the AI won't either.

Flawed Model Design can also introduce systemic bias, affecting the AI's decisions.

And don't forget Human Oversight—even the people building these systems can unintentionally perpetuate bias.

Chapter 11: Ethics and Responsible AI in Cybersecurity

When bias seeps into algorithms, the impacts can be serious.

There's the risk of Unfair Targeting, where certain groups may be disproportionately affected by security measures.

Overlooked Threats may slip by if algorithms are biased, weakening our defenses.

And this can lead to an Erosion of Trust—both users and organizations may lose confidence in cybersecurity solutions.

So how do we ensure fairness and reduce bias?

Start with Diverse Data Collection—using varied and representative data helps the AI perform fairly.

Carry out Bias Audits and Transparency checks regularly to ensure the AI's behavior stays aligned with expectations.

And adopt Inclusive Design principles to make sure the system is fair and accessible to all users.

When it comes to governance, ethical AI guidelines are key.

Chapter 11: Ethics and Responsible AI in Cybersecurity

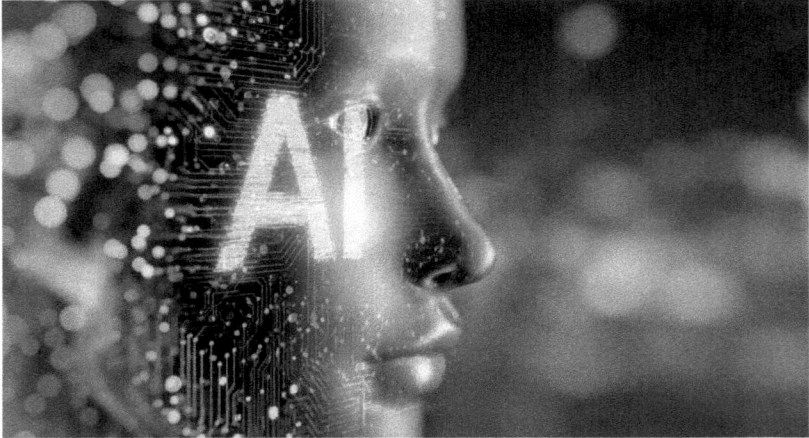

Ethical AI Guidelines promote responsible use across industries.

Risk Management in AI focuses on identifying and mitigating harms.

And Accountability in Cybersecurity is critical to ensure responsible handling of AI in sensitive security environments.

Let's look at some key regulations that affect AI in cybersecurity.

Data Privacy Requirements, like GDPR and CCPA, enforce strict rules around personal data handling.

Security Controls Enforcement ensures AI systems have robust safeguards in place.

And Transparency in AI Systems is essential for building accountability and trust with users.

To achieve compliance, organizations should focus on:

Chapter 11: Ethics and Responsible AI in Cybersecurity

Robust Data Governance—putting strong controls in place for managing data.

Regular Risk Assessments—constantly evaluating risks and adjusting practices.

Comprehensive Documentation—keeping detailed records for transparency and compliance.

And fostering Cross-Functional Collaboration—bringing together different teams to uphold AI ethics and ensure a unified approach to compliance.

Now let's discuss accountability in AI-powered cybersecurity systems.

Accountability means clearly defining who is responsible for the outcomes of AI decisions.

As AI increasingly automates cybersecurity, it's vital to make sure that responsibility for those decisions is clearly assigned.

Transparency and traceability are key to maintaining accountability.

Explainable AI helps users understand how decisions are made.

Detailed Logging keeps records of all system activities, aiding traceability.

And Audit Trails provide a chronological record, making it easier to investigate incidents and understand the AI's behavior.

So how do we address responsibility when AI systems fail or make errors?

Start with Clear Policy Frameworks that define accountability.

Have a solid Incident Response Plan in place for quick action when issues occur.

Legal Accountability ensures that there are clear consequences for failures.

And setting out Remedial Actions helps ensure that harm from AI errors can be mitigated effectively.

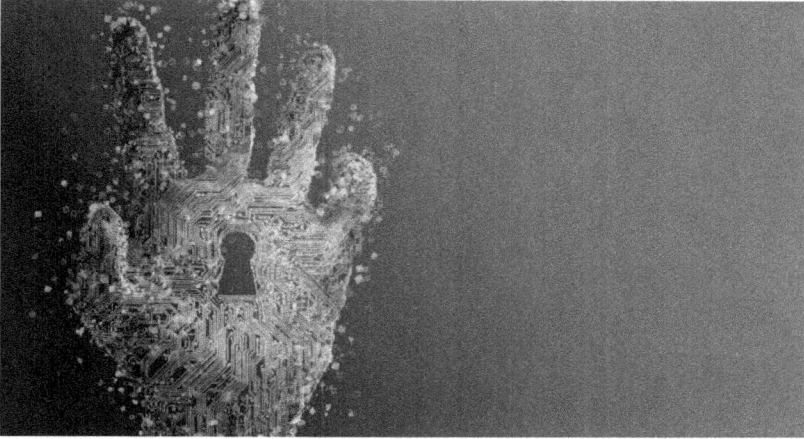

To wrap up, ethical and responsible AI in cybersecurity isn't just a technical issue—it's about fairness, transparency, governance, and accountability. By focusing on these areas, we can build AI systems that not only protect us from threats but also uphold trust, respect user rights, and comply with regulations.

Chapter 12: AI Red Teams and Blue Teams Collaboration

Let's start by defining the roles of Red Teams and Blue Teams in cybersecurity. Red Teams are like ethical hackers—they simulate cyber attacks to find vulnerabilities before real attackers do. Blue Teams, on the other hand, are the defenders. They monitor systems, detect threats, and respond quickly to attacks. When AI is integrated into both teams, it boosts their capabilities—helping Red Teams create smarter attack simulations and Blue Teams detect and respond to threats more effectively.

AI plays a critical role in enhancing the operations of both Red and Blue Teams. For Red Teams, AI can automate complex attack scenarios, making penetration tests more realistic and challenging. For Blue Teams, AI speeds up threat detection and response times, improving defense strategies. When both teams use AI together, it leads to more realistic tests and stronger security overall—a real win for cybersecurity.

AI-based threat simulations are designed to mimic real-world cyberattacks. These realistic scenarios help security teams test how well their defenses hold up against new and evolving threats. By practicing with these AI-driven simulations, teams can improve their ability to prevent breaches before they happen, making their defenses much stronger and more proactive.

There are powerful tools and platforms that create AI-driven cyberattack simulations. These platforms help teams evaluate both system vulnerabilities and how well their defenses respond. Because these environments are controlled and risk-free, they allow Red and Blue Teams to safely practice, refine their tactics, and improve their strategies without jeopardizing actual systems.

It's not enough to just run simulations—you also need to measure how effective your defenses are. Detection rates show how well threats are identified. Response times tell you how quickly your team can react. And system resilience measures how well your systems continue operating, even during an attack. Together, these metrics help teams fine-tune their security strategies.

Training human analysts with AI tools is key to modern cyber defense. These training programs teach analysts how different AI technologies work and where they're most effective. They also highlight the limitations of AI so analysts have realistic expectations. Most importantly, the training focuses on how humans and AI can work together to make smarter, faster decisions during a cyber incident.

Building strong decision-making processes between humans and AI is essential. Collaboration protocols help guide this interaction, making decisions faster and more accurate. While AI can process huge amounts of data quickly, human intuition adds critical context that AI might miss. When both work together, they form a powerful defense team.

Interactive co-training environments take learning to the next level. By combining AI tools with human judgment in training exercises, analysts sharpen their skills and improve situational awareness. These sessions encourage critical thinking and adaptability—traits that are crucial when facing constantly evolving cyber threats.

To keep improving, both AI systems and human teams need continuous feedback. Regular feedback loops help identify weak spots and fine-tune defense strategies. This approach allows defenses to adapt and evolve alongside emerging threats. Over time, collaboration between humans and AI leads to better performance and stronger cyber defenses.

Chapter 12: AI Red Teams and Blue Teams Collaboration

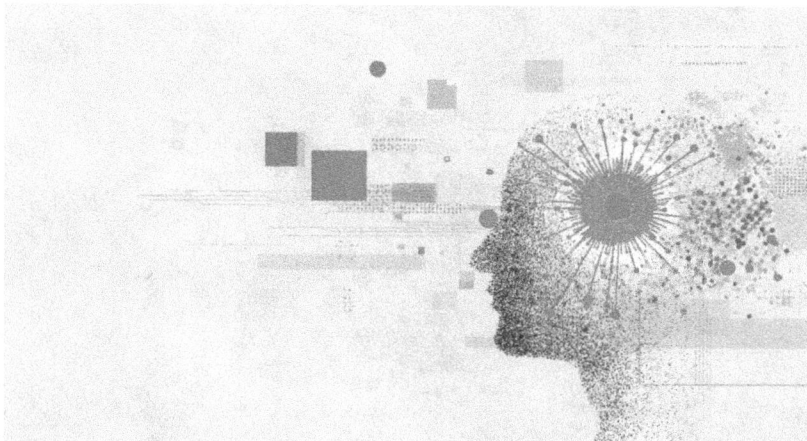

Cyber threats are always changing, so it's important to stay one step ahead. By studying past incidents, teams learn valuable lessons to prevent future attacks. Integrating threat intelligence into your defenses helps spot new risks early. And, by regularly simulating and testing your defenses, you keep your team ready for whatever comes next.

Building resilient defense systems is an ongoing effort. It starts with regular assessments to uncover new vulnerabilities. Continuous training keeps personnel sharp and ready to respond. And by embracing new technologies, defense systems stay robust and adaptable in a constantly changing cyber landscape.

To wrap up, collaboration between AI Red Teams and Blue Teams is key to building strong, adaptable cyber defenses. By simulating realistic threats, co-training analysts with AI, and committing to continuous learning, organizations can stay ahead of cyber attackers and protect their critical systems.

Chapter 13: Generative AI and Large Language Models

Generative AI refers to systems that can create new content—like text, images, or data—based on user inputs. Unlike older rule-based systems, generative AI uses deep learning to produce outputs that seem much more realistic and human-like. These models can create across different domains, whether it's writing, art, or design, showing just how versatile this technology has become.

Large language models are trained on massive datasets, allowing them to understand and replicate diverse language patterns. They use transformer architectures, which are especially good at handling complex language tasks. These models can understand context, perform zero-shot learning—where they handle tasks they weren't explicitly trained on—and adapt to a range of applications.

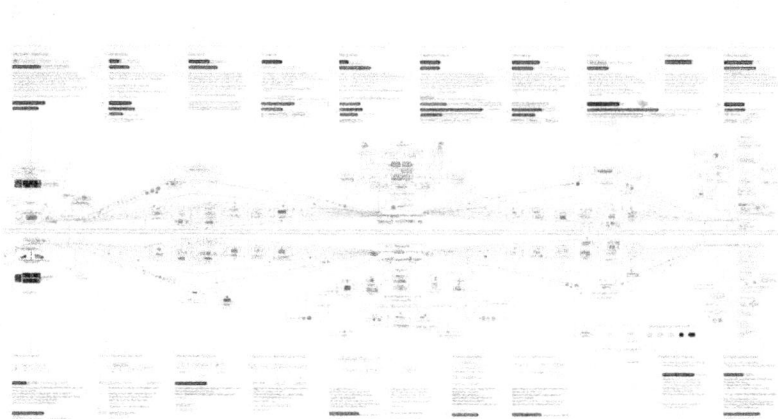

We're seeing rapid advancements in AI, especially with larger models that offer improved performance. Training techniques have also evolved, making models more efficient and accurate. An exciting trend is multimodal AI—where models can handle multiple types of data, like text and images. And importantly, we're now seeing these models deployed in real-world production environments.

AI can be a powerful tool for simulating sophisticated cyber threats. For example, it can generate realistic phishing emails to train security teams. It can also produce malicious code snippets for testing cybersecurity defenses. These simulations are valuable because they help organizations prepare for real-world threats, improving detection and response before an actual attack occurs.

AI doesn't just simulate threats—it can also create highly convincing phishing messages and social engineering attacks. By making manipulative content seem more authentic, AI increases the risk of successful scams and breaches. The realism provided by AI means organizations have to be extra vigilant against these kinds of attacks.

On the positive side, AI-driven content automation boosts efficiency and scalability, making it easier to create large volumes of content quickly. However, this same automation brings risks—particularly the spread of misinformation or fraudulent content, which can happen rapidly and at scale. It's a double-edged sword that needs careful management.

Prompt injection is when attackers manipulate input prompts to make a language model produce harmful or unintended outputs. Because these models tend to follow instructions very literally, it's possible for attackers to craft inputs that lead to undesirable behavior. This is a key vulnerability area for large language models.

There are several ways attackers exploit language models. Data poisoning involves tampering with training data to degrade a model's behavior. Adversarial prompts are crafted specifically to trick the model into giving harmful outputs. And sometimes, attackers exploit existing biases in the model to bypass safety measures. All of these tactics pose serious risks.

Chapter 13: Generative AI and Large Language Models

The impacts of these vulnerabilities are very real. Data breaches can happen if LLMs are exploited. They can also be used to spread misinformation, undermining trust in information sources. Over time, repeated misuse of these models can erode public confidence in AI systems and digital platforms altogether.

There are several technical strategies we can use to defend against these risks. Input sanitization helps prevent harmful data from reaching the model. Fine-tuning with safety constraints makes the model's outputs more reliable. Access controls ensure only authorized users can interact with sensitive systems. And continuous monitoring helps spot and address threats quickly.

Human oversight is a critical part of defense. Having a human in the loop improves detection of malicious outputs and allows for timely intervention when problems arise. By balancing automated tools with human judgment, organizations can create a more robust security posture against AI-related threats.

As we look to the future, deploying large language models securely will require a multi-layered approach—combining technical controls, human oversight, and evolving best practices. The goal is to maximize the benefits of AI while minimizing the risks, ensuring that these powerful tools are used safely and responsibly.

To wrap up, we've seen how generative AI and large language models offer incredible opportunities—but also come with serious risks. From simulating threats to being exploited themselves, LLMs need careful handling. By applying robust defenses, combining technical measures with human oversight, and staying informed about emerging trends, organizations can use AI securely and effectively.

Chapter 14: AI-Driven Risk Management

AI-driven risk management is about leveraging artificial intelligence to enhance how organizations handle risks. AI helps identify potential risks more quickly and accurately across various business scenarios. It improves the precision of risk assessments, allowing for better decision-making, and it empowers organizations to mitigate risks proactively by predicting and addressing issues before they escalate.

There are four key components to integrating AI in risk management processes. First, data collection and preprocessing—gathering and cleaning data is essential for accurate analysis. Next, machine learning models detect risks by identifying patterns in historical and real-time data. Automation plays a big role, handling repetitive tasks and reducing human error. Finally, continuous feedback loops allow AI systems to learn and improve over time, supporting better decision-making.

AI adoption in risk management brings significant benefits, including improved risk visibility by quickly analyzing large datasets and faster decision-making through timely insights. However, there are challenges too. AI requires high-quality data, and poor data can reduce effectiveness. Ethical issues, algorithm transparency, and the complexity of integrating AI into existing systems are also critical considerations organizations must address.

AI automates regulatory compliance by scanning large volumes of regulatory data continuously, ensuring organizations stay informed about relevant requirements. It maps organizational policies directly to these regulations, which helps maintain alignment and clarity. Importantly, AI can detect potential non-compliance issues early, allowing compliance teams to intervene before problems escalate.

One of AI's strengths is real-time data analysis. It can detect anomalies in data instantly, which may indicate compliance breaches. This immediate detection gives organizations the ability to respond quickly, potentially avoiding regulatory violations and the associated penalties.

Continuous AI-driven monitoring helps reduce operational risks by identifying potential problems before they disrupt business operations. Monitoring also ensures ongoing regulatory compliance, which reduces the risk of sanctions and financial penalties. Overall, early risk detection supports a more proactive and cost-effective approach to managing operational risks.

AI uses several methods to quantify cyber threats. Anomaly detection spots unusual patterns in systems and networks that might indicate a threat. Behavioral analytics examines user and system behavior to identify suspicious activity. Machine learning models analyze a wide range of data sources to score and assess cyber risks dynamically, allowing organizations to prioritize their security efforts effectively.

Integrating AI with cybersecurity frameworks enhances risk scoring and guides policy adjustments to improve security. AI insights help organizations adapt their policies dynamically and optimize incident response strategies, allowing for faster threat mitigation. Continuous AI monitoring also supports compliance by ensuring ongoing tracking and reporting of security-related risks.

AI plays a key role in supporting executive decision-making by providing data-driven insights. It helps leaders assess risks in uncertain situations, offering clear and actionable intelligence. With AI, executives can make more informed decisions by relying on comprehensive risk evaluations.

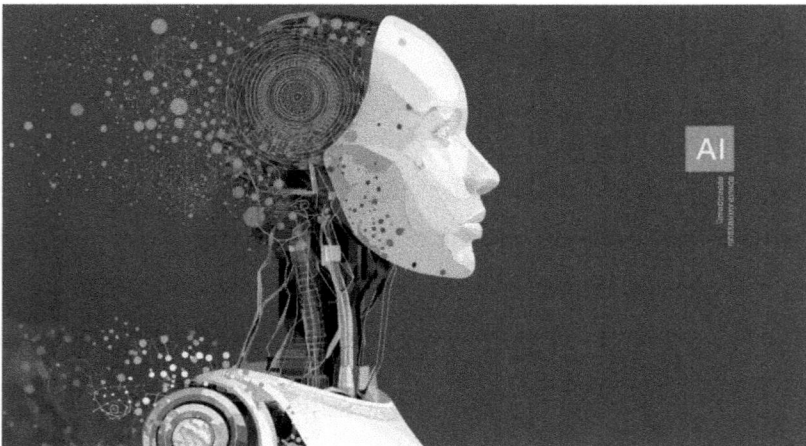

Chapter 14: AI-Driven Risk Management

AI-powered simulations allow organizations to analyze complex risk scenarios and anticipate potential outcomes. These simulations enable proactive planning and strategic decision-making. By understanding possible future risks, organizations can also develop effective mitigation strategies to reduce potential negative impacts.

AI boosts organizational agility and resilience by enabling rapid adaptation to emerging risks. Its continuous learning capability ensures that insights evolve alongside changing environments. With AI, organizations can keep their risk strategies updated dynamically, allowing them to respond more effectively to new threats and challenges.

In conclusion, AI-driven risk management empowers organizations to handle risks more proactively and intelligently. By enhancing risk visibility, improving decision-making, and supporting strategic planning, AI transforms how organizations approach both existing and emerging risks. However, realizing these benefits also requires addressing data quality, ethical concerns, and integration challenges.

Chapter 15: Building an AI-Enabled Security Program

An AI-enabled security program is built on three pillars. First, enhanced threat detection—AI helps detect threats faster and more accurately than traditional methods. Second, automated response—AI-driven automation speeds up incident response and reduces manual work. Third, improved security posture—by continuously learning from data, AI can strengthen the organization's overall security.

The strategic benefits of adopting AI in security are clear. AI enables improved threat identification, increasing accuracy and reducing risk. It leads to faster incident response, minimizing damage when security events occur. And with predictive analytics and scalability, organizations can proactively address risks and adapt to growing security demands.

When it comes to your security team, the right skills are essential. First, they need a solid understanding of machine learning algorithms to build effective AI solutions. They also need a strong foundation in cybersecurity fundamentals to apply AI responsibly. And don't forget AI ethics and data analytics—ensuring that AI is used ethically and that insights are drawn responsibly from data.

To build and maintain AI expertise, you need to foster a culture of continuous learning. Ongoing education—like workshops and seminars—keeps teams updated on AI advancements. Hands-on projects are crucial too; they give your team the chance to apply their knowledge in real-world scenarios, strengthening their skills over time.

Choosing the right AI vendors starts with setting clear criteria. Consider vendor experience and technology maturity—you want reliable solutions. Look at scalability and security to ensure the AI can grow with your needs and protect sensitive data. And always check for compliance and strong support, making sure the vendor meets regulatory requirements and offers dependable service.

Assessing vendor capabilities means digging deeper. Ask about AI model explainability—do they clearly explain how their AI works? Review their data privacy policies to protect your users. Check how frequently they update their AI models to stay current. And evaluate how open they are about the limitations of their AI solutions to ensure realistic expectations.

Vendor risk mitigation is critical. Start with security audits to uncover any weaknesses. Conduct reference checks to learn about their past performance. Always verify compliance with relevant regulations. And use contractual safeguards to clearly define responsibilities and reduce risks before signing any agreements.

When integrating AI with your existing security tools, look closely at compatibility. Evaluate API compatibility to ensure systems can connect properly. Check for data format consistency so information flows smoothly. And review operational workflow integration to avoid disrupting your existing processes.

Chapter 15: Building an AI-Enabled Security Program

Successful AI integration requires careful planning. Use a phased deployment approach to minimize disruption. Engage stakeholders early to ensure everyone is aligned. Perform thorough testing to catch issues before full rollout. And maintain clear documentation for reference and future maintenance.

AI systems aren't set-and-forget. You need continuous monitoring and optimization. Regularly evaluate performance to see if AI is delivering as expected. Set up feedback loops to help the system learn and improve. And don't forget model retraining and adaptation—update models with new data to stay effective against evolving threats.

To wrap up, building an AI-enabled security program means balancing opportunities with challenges. Focus on enhancing threat detection, improving response, and integrating AI responsibly. Invest in the right skills, carefully choose vendors, and plan integration thoroughly. Continuous monitoring and adaptation will help you maximize the benefits of AI in your security strategy.

Fraud detection is a top priority for financial institutions, and AI is transforming this area. AI algorithms can analyze transactions in real time, spotting suspicious activities and blocking fraud before it happens. Machine learning models get smarter over time, improving their ability to detect fraud while reducing false positives — meaning customers aren't unnecessarily blocked or flagged.

Threat intelligence is another critical use of AI in finance. AI systems bring together data from various sources, helping organizations identify potential threats early. By analyzing these patterns, AI supports proactive risk assessments, allowing banks and financial institutions to strengthen their defenses before an attack occurs.

AI also plays a key role in securing digital banking transactions. Machine learning models help verify user identities, ensuring only authorized users can access banking services. These models also monitor transactions continuously to detect unusual patterns, providing a seamless yet secure experience for customers.

In healthcare, protecting patient data is paramount. AI helps by detecting unusual access patterns that might indicate a data breach. It also enforces privacy policies by controlling access to sensitive data and ensuring compliance. Additionally, AI-enabled encryption methods keep patient information confidential and secure.

Connected medical devices introduce unique security challenges. AI systems monitor these devices for signs of abnormal behavior, which could signal a cyber threat. By detecting these issues early, AI helps prevent device exploitation — protecting both the device's functionality and, most importantly, patient safety.

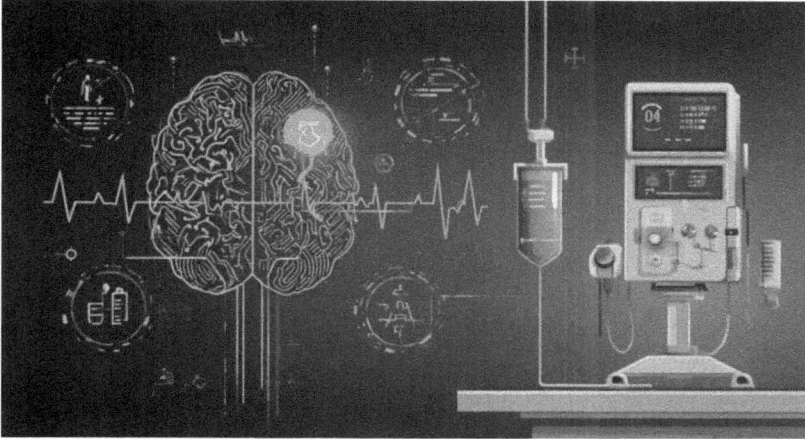

Compliance is a constant challenge in healthcare, and AI can ease the burden. AI tools track regulatory changes automatically, ensuring that organizations stay compliant without manual checks. This automation reduces the workload on compliance teams while minimizing the risk of regulatory penalties.

In industrial environments, real-time monitoring is crucial. AI-powered systems analyze sensor data from industrial control systems, spotting anomalies that could indicate system failures or cyber incidents. Early detection allows for quick responses, helping to maintain operational integrity and safety.

AI is also transforming predictive maintenance. By analyzing data patterns, AI forecasts equipment failures, allowing for timely maintenance and reduced downtime. In terms of cybersecurity, AI helps assess vulnerabilities, giving organizations a better chance of preventing cyber attacks before they occur.

Securing critical infrastructure is a high-stakes priority. AI-driven monitoring systems watch over critical assets — from power grids to water supplies — detecting and responding to cyber-physical threats in real time. These systems help prevent major disruptions and ensure essential services remain available.

In the public sector, AI enhances threat intelligence by analyzing vast amounts of data from different sources. This helps agencies identify threats early and act proactively. With AI, governments can strengthen their cyber defenses and stay ahead of constantly evolving adversaries.

AI also plays a key role in automating incident response. Automated detection tools monitor networks, identify threats, and isolate them before they spread. AI-driven mitigation tools help contain and respond to attacks quickly, reducing the impact on public sector organizations.

AI ensures continuous enforcement of cybersecurity policies, reducing the risk of human error. Intelligent systems also help organizations comply with regulatory requirements, maintaining governance standards and strengthening overall security posture.

To wrap up, AI is proving to be a powerful ally in cybersecurity across various sectors. Whether it's protecting financial transactions, securing medical devices, safeguarding industrial systems, or defending public sector networks, AI-driven solutions are enhancing security measures, improving threat detection, and enabling quicker responses to cyber threats.

Chapter 17: The Road Ahead

Artificial intelligence is playing a critical role in creating encryption methods that can resist quantum computing attacks. AI-driven algorithms help design encryption schemes that stay robust even against the power of quantum decryption. Post-quantum cryptography works as a safeguard, protecting sensitive data from future quantum threats. Additionally, AI helps optimize encryption methods, balancing stronger security with better performance.

Machine learning plays a vital role in detecting threats within quantum-enabled environments. These models analyze vast, complex datasets to uncover new vulnerabilities that traditional methods might miss. But quantum computing itself brings new, unique risks, so we need advanced detection techniques. With AI-driven detection, we can enable more proactive cyber defense strategies, staying ahead of increasingly sophisticated attacks.

Chapter 17: The Road Ahead

AI and quantum technologies each have unique strengths, and together they offer powerful advantages for cybersecurity. However, integrating these systems isn't without challenges — like managing computational complexity and overcoming technical barriers. Plus, as these technologies evolve, they bring new threat landscapes we must adapt to. That's why ongoing research and innovation are so important at this intersection.

Autonomous cybersecurity is built on key principles. First, self-learning capability — these systems improve over time by learning from new data without human input. Second, adaptive responses — they change and refine their strategies in real-time as threats evolve. And third, minimal human intervention — making them faster and more efficient at responding to cyber threats.

Machine learning enables real-time threat mitigation. These systems monitor data continuously, detecting threats as they happen. Once a threat is identified, the system can analyze it and respond autonomously, often before human operators even realize there's a problem. This greatly reduces response times and limits the damage potential.

Looking at real-world examples, autonomous defense systems have shown they can enhance threat detection accuracy, even in highly complex environments. They also improve operational resilience, keeping critical systems running smoothly under pressure or attack — a key benefit in today's cybersecurity landscape.

Governments worldwide are working on legislation to address cybersecurity risks posed by AI and quantum computing. These efforts aim to strike a balance between encouraging innovation and protecting security. New cybersecurity frameworks are being created specifically to tackle the challenges brought by emerging technologies.

International collaboration is essential in cybersecurity because threats don't respect national borders. Developing interoperable standards helps ensure different systems can work together seamlessly. Plus, sharing threat intelligence between countries and organizations allows for faster, more coordinated responses to global cyber threats.

Chapter 17: The Road Ahead

Policymakers face significant challenges with how quickly AI and quantum cybersecurity are evolving. They need to craft regulations that both encourage innovation and protect privacy. National security concerns must also be carefully managed — finding ways to safeguard against risks without hindering technological progress.

As we look to the future of AI and quantum cybersecurity, it's clear that continuous research, collaboration, and thoughtful policymaking will be critical. By embracing innovation while addressing emerging risks, we can build a more secure digital world that's prepared for the challenges ahead.

www.ingramcontent.com/pod-product-compliance
Lightning Source LLC
Chambersburg PA
CBHW060634210326
41520CB00010B/1605